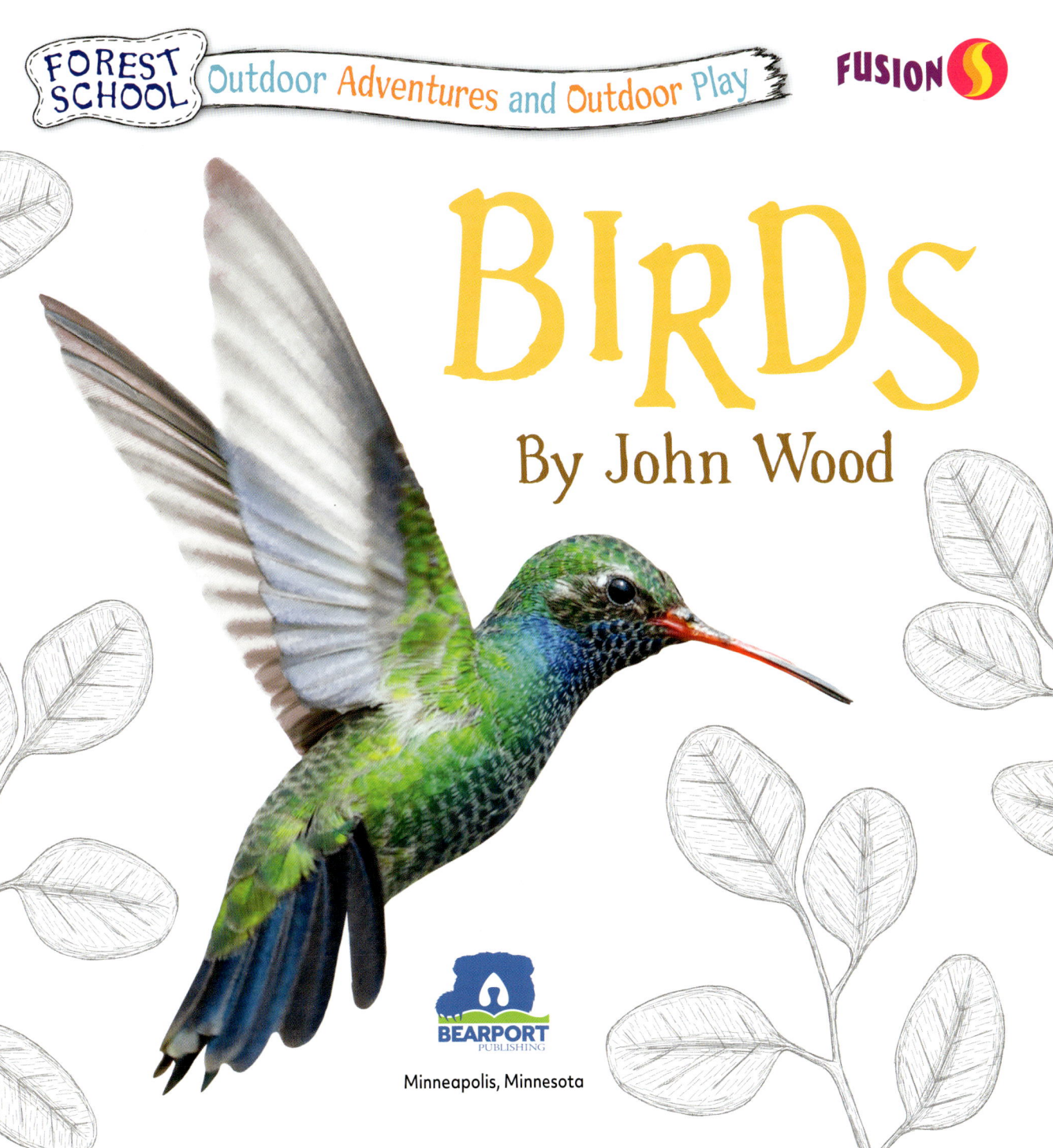

FOREST SCHOOL Outdoor Adventures and Outdoor Play

FUSION

Birds

By John Wood

BEARPORT PUBLISHING

Minneapolis, Minnesota

Credits:

Front Cover – Glass and Nature, myteria, ANURAK PONGPATIMET, 2&3 – Butterfly Hunter, 4&5 – Torychemistry, Denis Kuvaev, 6&7 – DNF Style, Odua Images, 8&9 – outtakes/iStock, Monkey Business Images, 10&11 – Evan Linnell/iStock, Bonnie Taylor Barry, 12&13 – Nature lapse, Vitalart, 14&15 – Tani_Bel, Hanna Taniukevich, 16&17 – TAUFIK ARDIANSYAH, Shaun Wilkinson, 18&19 – ps50ace/iStock, sasimoto, Roman Malanchuk, Bachkova Natalia, SeventyFour, 20&21 – ALEXANDER KOLIKOV, Rawpixel.com, Kristine Rad, 22&23 – Ann Worthy, Santhosh Varghese. Images are courtesy of Shutterstock.com. With thanks to Getty Images, Thinkstock Photo, and iStockphoto.

Library of Congress Cataloging-in-Publication Data is available at www.loc.gov or upon request from the publisher.

ISBN: 978-1-63691-461-9 (hardcover)
ISBN: 978-1-63691-468-8 (paperback)
ISBN: 978-1-63691-475-6 (ebook)

© 2022 Booklife Publishing
This edition is published by arrangement with Booklife Publishing.

North American adaptations © 2022 Bearport Publishing Company. All rights reserved. No part of this publication may be reproduced in whole or in part, stored in any retrieval system, or transmitted in any form or by any means, electronic, mechanical, photocopying, recording, or otherwise, without written permission from the publisher.

For more information, write to Bearport Publishing, 5357 Penn Avenue South, Minneapolis, MN 55419. Printed in the United States of America.

CONTENTS

Welcome to the Forest 4

Taking Care of Nature 6

Bird Watching 8

Feathers 10

Nests 12

Tracks 14

Bird Songs 16

Hungry Birds 18

Get Making! 20

Time to Think 22

Glossary 24

Index 24

Welcome to the Forest

Welcome to forest school. Let's explore, play, and create!

What do you want to learn about in the forest?

Get ready for forest fun!

We can learn so much from the world around us. Step outside into a great big classroom full of trees, birds, bugs, and more.

Taking Care of Nature

Any time we go into **nature**, we must take care of it. We should leave the forest as we found it.

The forest is home to many animals and plants.

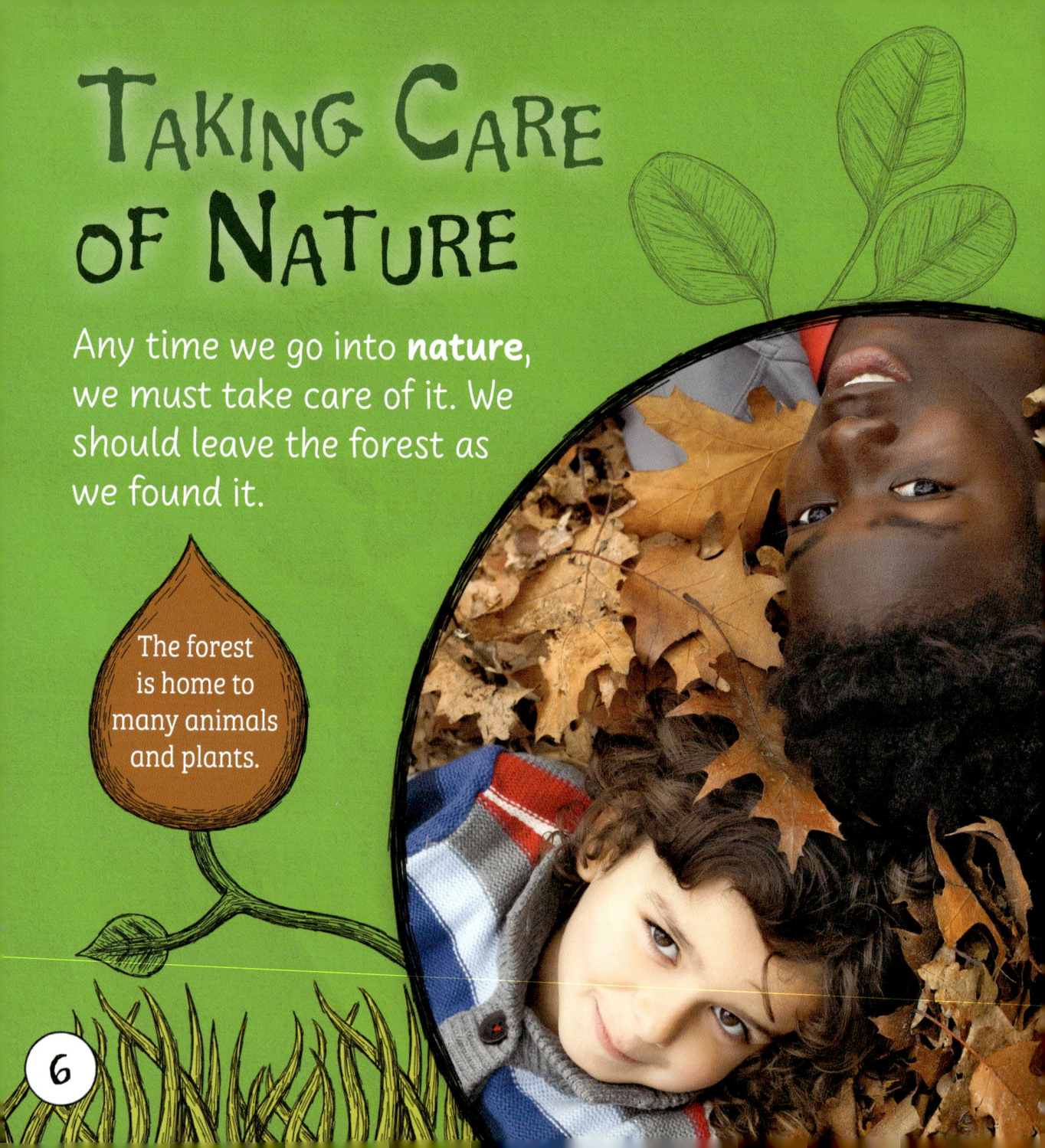

Stay on the path during forest school. That way, we won't hurt any plants or animals. What else can we do to care for the forest?

Let plants grow instead of picking them.

It's okay to watch birds and other animals. But remember not to touch them.

Take away trash so it doesn't become **litter**.

7

Bird Watching

Do you hear chirping? Look up and all around. Can you find a bird?

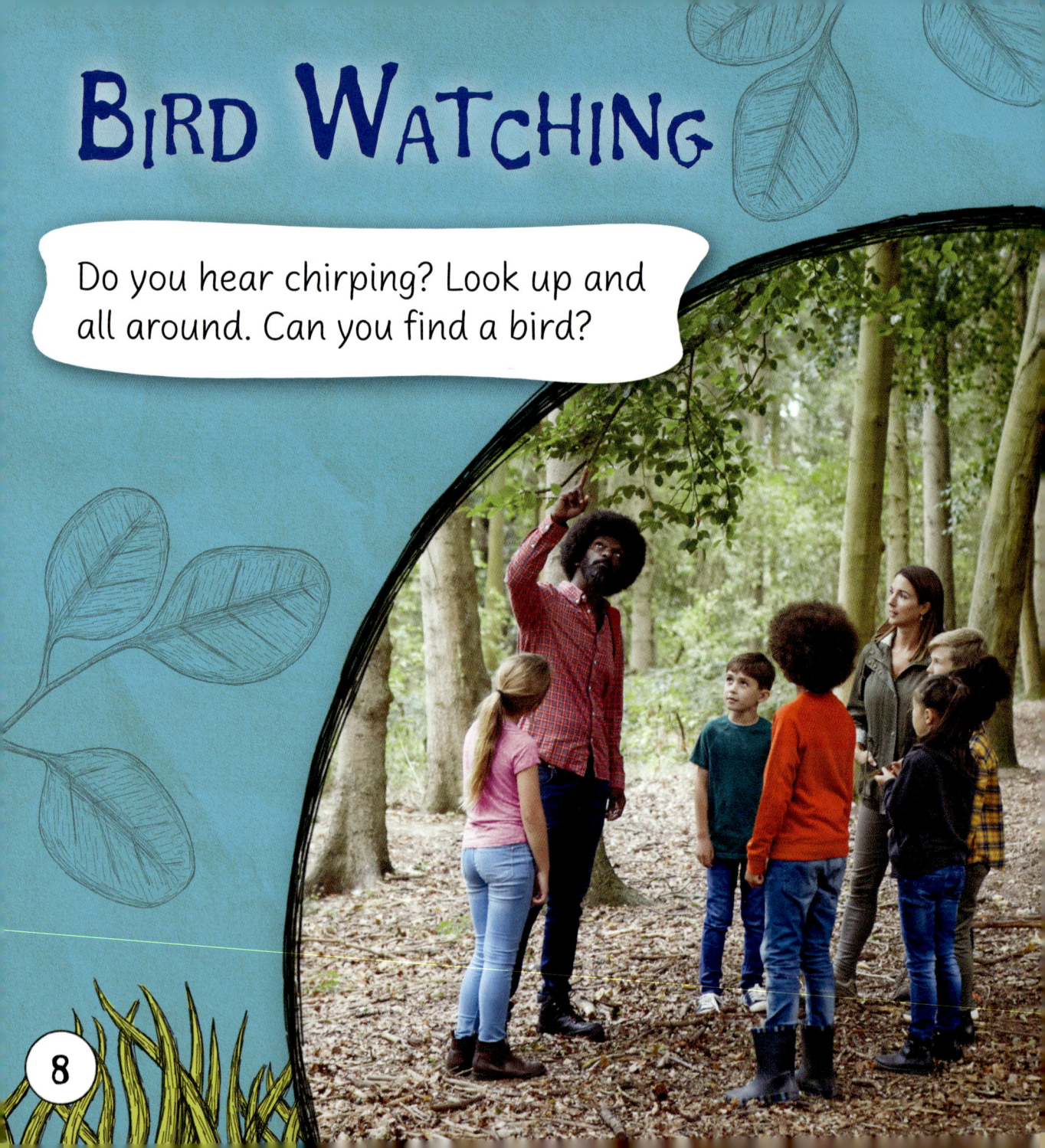

Birds can fly high in the sky. They sometimes sit on tree branches. Let's learn more about these amazing animals!

Many kinds of birds live in the forest.

FEATHERS

All birds have feathers. Feathers keep birds dry and warm. Many birds grow new feathers every year.

Some feathers have colors that help birds hide by blending in with the forest. Others have bright colors that help birds get one another's attention.

Male birds are often more brightly colored than **females**.

Nests

Birds gather sticks and leaves to build nests. They hold these things together with spider webs, mud, and even their own spit!

Birds fill their nests with soft things, such as feathers and fur.

Be sure to stay back when you look at nests. Give the birds some space.

Nests are safe places for birds to lay eggs and raise babies.

Tracks

Birds leave tracks when they walk on the ground. The shape of the tracks can help you guess what kind of bird left them.

Ducks have **webbing** between their toes. You can see the webbed shape in their tracks.

Tracks from eagles and hawks show three toes pointing forward and one pointing backward.

You can look for tracks in snow, dirt, or sand.

Bird Songs

Can you hear the birds singing? It's more than just a pretty song. Birds sing to **protect** their homes and tell other animals to stay away.

Male birds also sing to show off to females. Birds often sing more in the morning.

Some birds copy other animal sounds, such as a frog's croak or a cat's meow!

Hungry Birds

Have you ever seen a bird eating? Many birds eat fruits, seeds, nuts, bugs, and worms.

Seed

Worm

Fruit

Some birds hunt and kill their food. They may eat fish, small animals, snakes, or even other birds!

Sometimes, birds eat animals that have already died.

Get Making!

Want to take your bird watching to the next level? Make a list of all the birds you spot in the forest.

You could also make a **recycled** bird feeder out of a plastic bottle. Carefully cut a hole in the side of the bottle so birds can reach inside.

Then, put **birdseed** in the bottle. Use string to hang your feeder from a tree branch.

Only fill your feeder with birdseed. Birds should not eat human food.

Time to Think

Our time at forest school is almost over. Let's think about everything we've learned.

Nature can be a great place to go and think.

22

Glossary

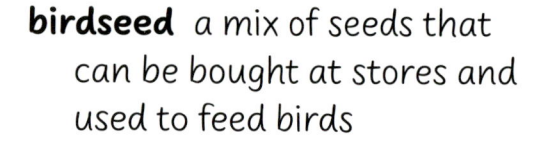

birdseed a mix of seeds that can be bought at stores and used to feed birds

females birds that can lay eggs to have baby birds

litter things that have been thrown away and are lying on the ground

male a bird that cannot have baby birds

nature the world and everything in it that is not made by people

protect to keep something safe

recycled used again to make something else

webbing skin between an animal's toes that helps with swimming

Index

eggs 13
feathers 10–12
fruits 18
nests 12–13
singing 16–17
sticks 12
tracks 14–15
trees 5, 9, 21
worms 18